TEN HOUSES
Edited by Oscar Riera Ojeda

世界小住宅 3

Peter L. Gluck and Partners

邵 磊 译

中国建筑工业出版社

图字：01-2000-0373 号

图书在版编目（CIP）数据

世界小住宅.3/（美）奥赫达编；邵磊译.—北京：中国建筑工业出版社，2000
（世界小住宅设计经典译丛）
ISBN 7-112-04256-9

Ⅰ.世... Ⅱ.①奥... ②邵... Ⅲ.①住宅－建筑设计－图集②住宅－室内装修－图集 Ⅳ.TU241-64

中国版本图书馆 CIP 数据核字（2000）第 21023 号

Copyright © 1997 by Rockport Publishers, Inc.

All rights reserved. No part of this book may be reproduced in any form without written permission of the copyright owners. All images in this book have been reproduced with the knowledge and prior consent of the artists concerned and no responsibility is accepted by producer, publisher, or printer for any infringement of copyright or otherwise, arising from the contents of this publication. Every effort has been made to ensure that credits accurately comply with information supplied.

本套图书由美国 Rockport 出版公司授权我社在中国翻译、出版、发行该套图书中文版

责任编辑：张惠珍　程素荣
美术编辑：黄　燕　姜敬丽

世界小住宅设计经典译丛
世 界 小 住 宅 3

彼得·L·格卢克合伙人事务所
奥斯卡·列拉·奥赫达　编
邵　磊　译

中国建筑工业出版社 出版、发行（北京西郊百万庄）
新 华 书 店 经 销
利丰雅高印刷（深圳）有限公司印刷
开本：255mm×230mm
2000 年 6 月第一版　2000 年 6 月第一次印刷
定价：**68.00** 元
ISBN 7-112-04256-9
TU·3354(9713)

版权所有　翻印必究
如有印装质量问题，可寄本社退换
（邮政编码 100037）

Contents 目　录

6	**序**	奥斯卡·列拉·奥赫达
8	**导言**	保罗·戈德伯格

精选作品

12	三个山墙住宅，康涅狄格州，莱克维尔
20	马诺尔音乐庄园，纽约，马马罗内克
30	双面住宅，马萨诸塞州，林肯
36	密斯住宅的亭和二期扩建，康涅狄格州，韦斯顿
50	"美国风"住宅扩建，纽约，普莱森特维尔
60	带游泳池和下沉花园的农宅，纽约，沃切斯特
72	直线型住宅，纽约，米勒顿
80	桥式住宅，纽约，奥利夫布里奇
92	有内院的湖畔住宅，伊利诺伊州，海兰帕克
100	城市住宅，纽约，布鲁克林

106	**精选建筑与项目简介**
107	**公司概况**
108	**照片提供者**

Foreword 序

by Oscar Riera Ojeda

Bridge House, Olive Bridge, New York

这本作品集给人的第一印象可能是在选材上大相径庭，在观点上也不尽一致。但是，这些作品的确出自同一位建筑师，他就是彼得·L·格卢克。他反对那种以浅显易懂与识别简单为特点的传统的标签手法。和同时代一些以风格模仿的特点的知名建筑师的作品相比，其自信而富有特色的手法创造了匠心独具的效果。这些都在本书的作品中得以清晰地体现。

在以喧闹和迅速的品名包装为特色的世界中，他的宁静而又前后截然不同的作品中蕴含的复杂性，绝非孤立地从表面上可以认识。我们必须跟随着这种不协调性，曲径通幽地去理解细节的精妙及其深度，理解针对建筑和设计的要求而采取的睿智方法，然后才能领略到在不同的表达之外，赋予作品生命的强烈而普遍的立场。

他的建筑的成功在于辩证的思想，其中最为普遍的是人对住宅的实在体验（它的设计和使用）与抽象的建筑形式之间的对立统一。在东方与西方、室内与室外、前方与后方、建筑与雕塑、自然与人造景观以及其他普遍遇到但缺乏审问的建筑现实条件之间，格卢克的作品以询问的姿态对其辨证关系进行了探究。

如果将这10栋住宅作为一组，分析其主要结构、细部和手工艺式的施工，我们会获得结束一种建筑困惑的享受。我们既可以看到他作为现代主义者的信仰，又可以感受到他那以更为自由与包容的态度，解放和扩展现代主义原则的决心。在这十个方案中，他在现代主义道路上，热烈地追求完整而且更新的地平线，在三个维度上引导着自己的创造性探索，用他的话说就是"戴面具"的现代主义、"语境化"的现代主义和"直率"的现代主义。本书开头的三个坡屋顶住宅是"戴面具"的现代主义的代表，格卢克运用了"住宅"最为人熟悉的解读方式，但寓于完整的现代主义形式之中。对于"语境化"的现代主义，三所现存住宅的加建设计方案是当之无愧的代表，他在设计中尊重现有建筑的语境，但坚持以"多重解读"而非复制的方式对待新老建筑。本书的最后四个作品则代表了"直率"的现代主义。他以需要单一解读的手法，运用抽象的形体，揭示了全新的更为包容的现代主义的可能性。

这本彼得·L·格卢克的住宅汇编，强调了为建筑的解决方案进行的多面研究，当然，建筑的回答永远不只一个。

奥斯卡·列拉·奥赫达

Introduction 导 言

by Paul Goldberger

彼得·L·格卢克是崇拜埃德温·勒琴斯(Edwin Lutyens)的现代主义者,这个有趣的事实或许是最富揭示性的。它告诉我们格卢克在某种意义上是个反对偶像崇拜者,因为勒琴斯毕竟是英国曾经有过的最伟大的古典主义者之一,有谁能想象一位强烈倾向现代主义的美国建筑师以他作为偶像呢?不过,勒琴斯不是书本上的古典主义者,他的建筑是灵活的舞蹈,是一种创造而非模仿。在勒琴斯那里,古典主义是一种语言而非意识形态,这正是格卢克对待现代主义的态度。格卢克醉心于对复杂问题富有想象力的解答,统一他的作品的不是外观的相似,而是一种前后一致的目的、睿智的探询、形式空间要素在感官与理智上产生的愉悦。

没有标准化的彼得·L·格卢克式建筑,它们总是由于位置、地段和业主的不同而变化。但纵览他的所有作品,无论是物质空间形式还是其中蕴含的哲理,都表达得畅快淋漓。格卢克的建筑不柔和也不多愁善感,相反,对于一些建筑师运用的历史主义的更为活泼的形体,格卢克却努力赋予其更为硬朗的轮廓,从而获得一种舒适,甚至是一种修养,同时也达到和环境无间的融合。格卢克以后现代主义的价值系统创造了现代主义的建筑,即建筑不是纯粹和抽象的雕塑,其设计和物质形态充满了对周围环境的回应。

格卢克设计的几个扩建工程最为有名,绝非偶然。位于纽约郊区的联邦风格(Federal Style)农宅、康涅狄格州城郊的密斯住宅以及位于韦斯特切斯特县"美国风(Usonian)[①]社区的赖特风格住宅,没有一个易于扩建,

---

[①] 借用萨缪尔·巴特勒(Samuel Butler)的小说《Erewhon》中的用法,代替 America——译者注

很难再想象比它们更加自成一体的建筑了。如何扩建而又不破坏其整体性?格卢克摒弃了复制的方式,在不同的方案中他的手法有着明显的不同。对联邦风格的农宅,他以长长的低矮的复杂几何体进行扩展,玻璃、石材、金属和木材,构成了精细的平衡,形成了果断、有差别的组合。为密斯住宅新建一个侧翼是一个更加微妙的难题,格卢克没有沿用农宅扩建的对比手法,而是对密斯的形式进行巧妙的重复,以细部的变化表现出鲜明的差别,区别于复制的方式,但不失对原建筑整体感觉和体量的尊重。赖特风格的住宅由赖特的一个学徒设计,位于陡峭的山坡之上。格卢克在原建筑下部设计了圆型的水泥基座,体现了赖特对圆形的偏爱。尽管这个设计和密斯住宅扩建的方案相比,在模仿风格的程度上不如后者那么字斟句酌,但它是对原建筑的美学精神和感觉的回应,而不是直接地模仿。

所有这些住宅都清楚地展示出,格卢克是一位建筑组合与细部设计并重的建筑师。尽管他的作品都十分精细,但他也同样胜任大胆直率的作风。詹姆斯·盖博勒·罗杰斯(James Gamble Rogers)在耶鲁达文波特学院设计中,将哥特风格和乔治风格的立面分置于两侧,格卢克位于麻省林肯郡的双面住宅应当之无愧地成为第二个同一建筑两种立面的例子(不知是否因为格卢克毕业于耶鲁大学建筑学院而受到影响?)。尽管将格卢克同罗杰斯相比稍有偏颇,因为耶鲁的建筑就是以极为断然的方式变化风格的,但无论如何,格卢克在这个设计中最为成功之处在于其微妙的渐变处理。沿街的前立面是抽象的山墙形式,体现出林肯郡的建筑环境与业主对私密性的要求。在后部,建筑变成了两层的、光滑的现代主义作品,面向河流、草坪和树林形成的大地景观。但最有趣的部分还在于两侧,可以看到坡屋顶的侧面被山墙截断。格卢克将这两个立面交织起来,迫使两种建筑语言从对立转向对话。

从一种真切的体验出发，强制性的对话是彼得·L·格卢克作品的主题，也是容易为人忽视的一点。密歇根湖畔住宅同时体现了对赖特的第二栋赫尔波特·B·雅各布住宅和国际风格的敬意，也表现了对几何形体吸引力的普遍偏爱，不过在这栋建筑中，几何体看起来一点也不武断生硬。住宅的形体组合仿佛有点不协调，但格卢克着意赋予其一种理性而非建筑的奇思妙想。强制性对话的实际存在赋予我们的感受，正是这件作品最有力度的方面。

是什么若隐若现、而又对格卢克的作品产生最重要的影响？答案是日本建筑，因为在70年代早期他在日本生活和工作了数年。在豪华的马诺尔音乐庄园（其山墙的立面和勒琴斯设计的巴顿·圣玛丽住宅有很大的类似）设计中，铜制的落水管、石灰岩石盆和柚木的灯具构件等细部，都体现了日本建筑的影响。联邦风格的农宅和格卢克自己的乡间住宅（一栋改造过的农宅，以一系列抽象的附属建筑，柔和地表达了家庭和乡村的本质）也不例外，比如其巨大的木制桁架。日本的影响还体现于更为潜在的方面，即一种特别的敏感性，我们不妨称之为"能动的敏感性(enabling sensibility)"，因为在格卢克复杂而且常常自相矛盾的形式之外，还存在着一种统一。通常而言，西方的建筑的宁静氛围和其简洁成正比，愈简洁就愈祥和。相反，日本的建筑是寓于高度的积极性中的宁静。格卢克就是从中汲取了灵感，运用在完全西方化的建筑语境中。他的建筑是积极的、复杂的，而从来不失宁静。

格卢克的作品中包含着各种奇思妙想，但不会让人觉得滑稽。相反，他的建筑在这个方面过于严肃，但是对格卢克而言，勿需千方百计而获得

灵感的方式，令人想起他对历史主义时隐时现的暗喻。他看到了是什么，明白了为什么，但他主要的业务目的是做一些其他的东西。作品在过程中看起来什么都像，但最终自成一体，不只是出于形式与空间的历史性、语境、现代主义、传统主义或者纯净性，而是当建筑师关注问题的所有方面，宣布它们和平共处时，生成的最终结晶。

保罗·戈德伯格

保罗·戈德伯格(Paul Goldberger)是纽约时报首席文化专题记者，从1973年起在该报主持建筑评论，并撰写了几本和建筑相关的著作，如《摩天楼及其进展：后现代时期的建筑与设计》。他在建筑学、建筑设计、城市规划和历史保护等方面涉猎很广。1984年获得普利策(Pulizer Prize)建筑评论奖。

Three-Gable House
Lakeville, Connecticut

三个山墙住宅
康涅狄格州，莱克维尔

Above: Retaining walls support bermed earth to create the proverbial suburban driveway-plus-basketball-court. The standard two-car garage door is hidden from the entrance, main level, and lawns, and the usually ad-hoc play space is given formal clarity and architectural definition.

Opposite Page: The three gables, their traditional form expressed as sculptural abstraction, present the house to the country road, the lake view, and the Connecticut hills. The second-floor overhang shields the first-floor windows from the southwest sun, and the colonnade beneath it jumps the scale of the facade, enhancing its formal purity, and also conceals the first floor's inexpensive standard windows and their conventional programmatic placement.

上图：挡土墙围合出典型的城郊式车道和篮球场。入口、主要楼层和草坪将双车型车库门隐藏起来，经常专门用作娱乐的场地被赋予形式的确定性和建筑的定义。

左页图：三个山墙的传统形式通过雕塑性的抽象进行表达，将住宅融入乡间道路和湖光山色之中。第二层的出檐起到了遮阳的作用，柱廊调整了立面的尺度，增加了其形式的纯净性，同时遮挡住首层价廉的普通窗户及其传统的设计布局。

此设计运用了价廉物美的建筑材料和传统的结构形式，以最为经济的建筑形式获得了尽可能多的房间。其业主是典型的建筑商，住宅在种种约束之下体现了优雅而且完整的居住性。

坡屋顶是住宅的主题。如同儿童绘画中的"家园"，单个的山墙、正面的门、烟囱和坡屋顶，住宅强调了美国传统的"家"的观念。独户住宅给人的长长的建筑意象中，舒适而又熟悉的坡屋顶并不排斥雕塑与形体的现代主义阐释。在这个方案中，夸张的坡屋顶不仅为住宅遮风避雨，无论是室内还是室外，其雕塑般的形式浑然一体。两个深陷的窗户强调了外观的体积感，陡峭的屋面使得室内屋顶空间得以更好地利用。

规矩的形体组织体现了基座、柱子、柱头等要素的古典性。首层部分地位于山坡的阶梯之下，成为在二层支撑着坡屋顶的内凹式柱廊的底座。这栋住宅不足5000平方英尺（约465m²），包括宽大的两层高的起居厅、厨房、主卧套间、四间次卧室、三间浴室、化妆间、洗衣间、储藏间以及双车型车库。

建筑只使用了两种基本的材料——沥青木瓦（asphalt shingles）和丙烯酸灰泥（acrylic stucco），因此造价低廉。一些细部诸如柱廊，将普通的木窗遮挡起来，同时给原本普通的设计与结构增添了雄伟。住宅内部十分宽敞，从远处看来，其雕塑性的外观远远超出了作为住宅所具有的意义。

Three-Gable House
三个山墙住宅

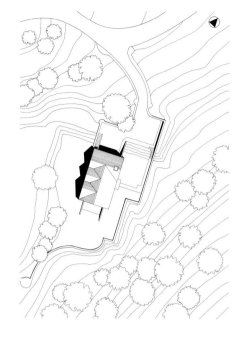

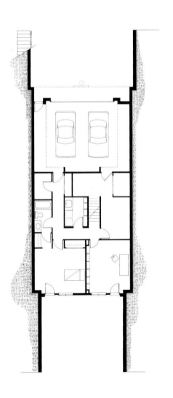

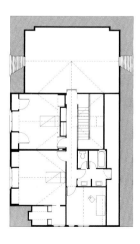

Three-Gable House
三个山墙住宅

上图：北立面上单个的厚重山墙标志出住宅的正式入口。和另一侧的三个山墙相比，这种传统的形式更加抽象。

下图：最初设计的车库后来改作娱乐活动室。在山上另建了一间独立式车库，也采用了山墙的形式，和住宅主体遥相呼应。

Above: *A strong single gable on the north facade marks the formal entrance to the "house"; here, the traditional image is even more abstracted than in the three gables opposite.*

Left: *The original garage was later converted into a playroom, and a detached garage was built into the hill, its gabled form echoing and complementing the larger composition.*

Three-Gable House
三个山墙住宅

右图：选择坡屋顶在于其象征的作用和低廉的造价，也增加了室内形式和空间的表现力。图中，二层的双面走廊打破了两层高的双坡式的起居空间。

右页图：普通的设计元素，比如楼梯和书架，成为表现雕塑感和展现光线与形体的舞台。

Right: The pitched roofs, chosen for their evocative symbolism and their low cost, also provide the occasion for expressive interior forms and spaces. Here, the typical double-loaded, second-floor hallway breaks through to the two-story, peaked living room.

Opposite Page: Ordinary residential elements—such as stairways and bookshelves—become opportunities for sculptural expression and for the play of light and mass.

Three-Gable House
三个山墙住宅

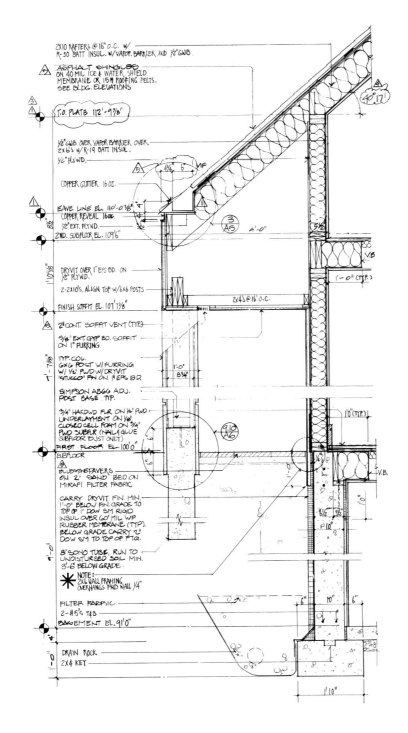

Manor House with Music

Mamaroneck, New York

马诺尔音乐庄园

纽约，马马罗内克

Above: A pool and trellised terraces create a foreground to views of Long Island Sound.

Opposite Page: A continuous wall provides a unifying enclosure to the entrance court. The two-story facade is rotated slightly, reducing its scale and subverting its symmetry.

上图：游泳池和格子花架形成了长岛海峡的前景。

左页图：连续的院墙形成了入口院落的完整围合。两层的立面相互之间略有旋转，减小了其尺度，并打破了对称。

这栋私人住宅中，设计了经常性的室内音乐表演空间和相当多的娱乐设施。尽管时常有孩子和客人拜访，但它仍然能够很好的满足一对夫妇的居住要求。为了既能举办100人的公共音乐会，又能专门进行面对两个人的表演，建筑必须根据实际的需要，可以非常宽敞，又能达到十分私密。

地段面对着长岛海峡，两侧都有建筑。住宅呈L形，将服务设施如洗衣、仆人用房、车库等和起居空间与公共区域分离。服务用房将住宅同与之毗邻的建筑隔离开来，起居用房则面向海岸，每个房间中都可以欣赏到清晰的海景。正式的入口与来访时的停车都在前院解决，四周的院墙上开有缺口，引入了周围的景观。院子在平面上稍有偏转，形成了不对称的布局，削弱了其规整的感觉，调整了立面的尺度。

窗户以不间断的韵律排列，对端头山墙的强调舒缓了理性主义立面的形式规律。在一系列的坡屋顶住宅中，此设计以屋顶作为众人熟悉的符号，生成现代主义风格的面具，以更好地和这个郊区社区的风格取得协调。

景观元素在两个方面积极地扩展了住宅空间：院落使之成为室外的公共空间，格子花架使游泳池成为室外面向海峡的私密区域。一系列相互关联的房间都被组织在一个更大的空间之中，同时还有一些过渡空间，如阳台，也不失为欣赏音乐会的好地方。住宅中没有尺度过大的单个房间，但当这些小房间结合在一起使用时，就能够为很多人提供服务。主人卧室套间在功能上则是相对独立的。

22　Manor House with Music
马诺尔音乐住宅

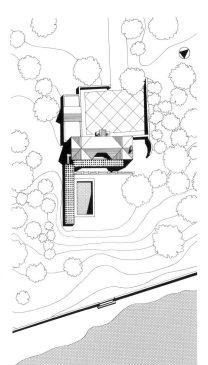

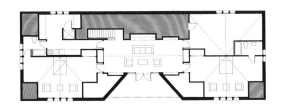
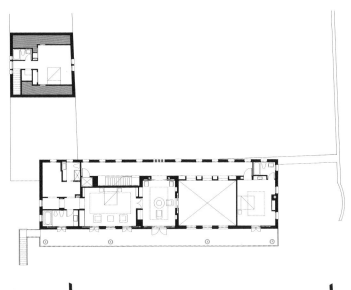
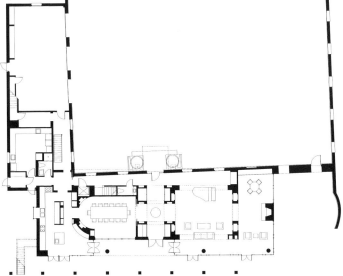

Manor House with Music   23
马诺尔音乐住宅

左图：每个房间都朝向水景，均有通向室外阳台的出口。室外阳台分隔了立面，减小了立面的尺度感，同时也成为遮风避雨的室外空间。

左页图：悬挑的混凝土楼梯从主卧和客人房间直接通向游泳池和院子。格子花架既汇集起面向海峡的景观，又为游泳池遮荫，同时还将其和西部的邻居分开。

Left: *Every room faces the water, each with access to an outside terrace or deck. The decks are cut into the facade to reduce its scale and to provide protected outdoor space.*

Opposite Page: *A cantilevered concrete stair gives direct access to the pool and grounds from the master bedroom and guest room. The trellis both focuses the view toward the Sound and provides shade for the pool while screening it from the neighbors to the west.*

26 | Manor House with Music
马诺尔音乐住宅

右图：在排水槽、落水管、雨罩等铜和石材的构件的装点下，总共三层的立面的规整形式变得活泼与亲切。雨水汇集到石盆之中，形成间歇喷泉。

右页图：方窗以富有理性的方式排列，和传统的坡屋顶形式的立面形成对比；入口和钢制的门罩、排水管以及中央的格子窗形成效果强烈的组合。这一切创造了鲜明的现代主义风格的意象。

Right: The imposing formality of the three-story elevation is enlivened and brought to human proportion by the sculptural play of copper and stone elements: gutter, downspouts, and entrance canopy. Rain fills the two limestone bowls, making intermittent fountains.

Opposite Page: In contrast to a traditional reading of the pitched-roof facade, a strong modernist image is created by the rationalist pattern of square windows and the strong entry figure with its steel canopy, drain spout assemblage, and central gridded window.

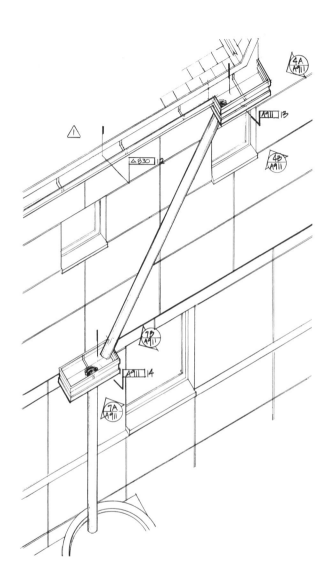

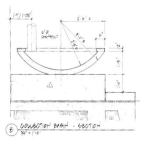
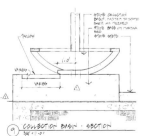

28 Manor House with Music
马诺尔音乐住宅

Right: The two-story living room—in plan, a room within a room—doubles as a chamber-music hall. Designed for acoustic quality, it seats one hundred listeners.

右图：从平面上看，两层高的起居室套在另一个房间之中，经过专门的声学设计，可以当作室内音乐厅使用，容纳一百名听众。

Manor House with Music

马诺尔音乐住宅

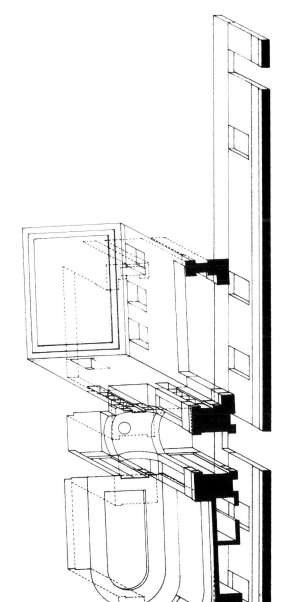

左图：一层平面呈直线形，在入口的轴线十字交叉处形成中厅和三角拱，整个一层都用作大型聚会的场所。楼上属于私密空间，主楼梯和公共空间相分离。从亲切的家庭起居室中可以鸟瞰院子，有一圆孔和下方的中厅相通。

Left: *The linear first-floor plan, with its central hall and pendentive arch at the entrance cross-axis, is designed to accommodate large gatherings. The upstairs is private, its major stairway shielded from the public spaces. The intimate family room looks out on the courtyard and down through an oculus to the central hall below.*

Two-Sided House
Lincoln, Massachusetts

双面住宅

马萨诸塞州，林肯

Above: From the town road, the house appears across the Sudbury River, childlike in simplicity, suggesting the familiar outlines of traditional New England architecture.

Opposite Page: From the rear, the house is unabashedly modernist. The south facade, which looks on to an unspoiled pond and bird sanctuary, is almost entirely glass. Second-story bedroom windows have a freestanding screen for awnings, while the overhang of the balcony above shades the first-floor kitchen and study.

这栋住宅的设计解决了诸多冲突：首先是城市远郊社区的保守风格和业主对现代主义风格的钟爱之间的矛盾；其次，业主希望建筑在社区之中不显山露水而在家中应个性十足；第三，建筑开放、通透的同时还得满足对温馨、私密的围合空间的需要。

因此，住宅有着不同的两面。在面向公众的一边，立面的山墙和檐口唤起对传统的新英格兰建筑的记忆，但又没有完全的复制。巨大的坡屋顶上没有设置窗户，有效地阻挡了北风的侵袭，亦保护了住宅的前门。在私密的一边，南立面上有大面积的玻璃，面向太阳和萨德伯里河两岸。其白色的立面充分地体现了现代主义的形式和表达的抽象性。

分列式的外观在住宅侧面结合，在这里，传统的围合语言和现代的开放性相遇。住宅室内是完全现代的，透过窗户和天窗的光线在空间中跳动。在平面上，室内的公共空间完全开放，并围合出室内花园。二层儿童卧室旁侧的走廊有房间大小，在坡屋顶檐下的位置上，设置了像床那么大的玻璃窗式座位，形成了所谓的"发现空间"，专门用来娱乐和接待特别的客人。总之，这栋建筑是家庭住宅，有着很好的适应性。

上图： 从乡间公路上望去，住宅位于萨德伯里河一侧，以孩子般的单纯，展示出传统新英格兰建筑的轮廓。

左页图： 从后部看，住宅落落大方地展示着其现代风格。南立面几乎全部由玻璃组成，面向得到妥善保护的池塘和鸟禽保护区。二层卧室的窗户有遮阳帘，悬挑的阳台对一层的厨房和学习间起到了遮阳的作用。

Two-Sided House
双面住宅

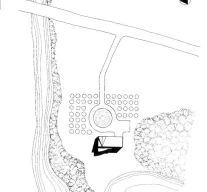
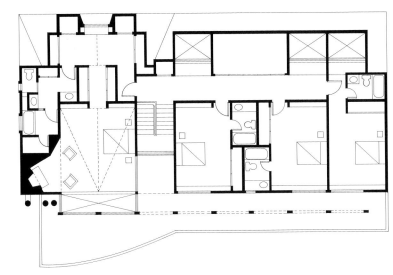

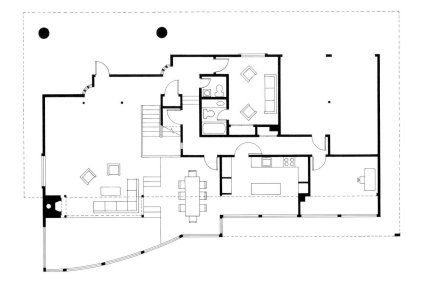

Two-Sided House 33
双面住宅

左页图和左图： 截然不同的立面在住宅的侧面相结合，坡屋顶和纯粹垂直的形体连接在一起。从街道上看，住宅唤起了对传统乡土风格的记忆，不过并不是模仿得惟妙惟肖：入口处的山墙和支柱体量过大；入口的悬挑也过于夸张，好像浮在空中没有什么支撑；从近处看壁炉烟囱，它似乎漂浮在坡屋顶上。和北立面形成对比，南立面摒弃了所有的传统形式。

Opposite Page and Left: Resolution of the two opposing facades occurs on the side elevation, where the pitched-roof forms interlock with the pure orthogonal shapes. From the street, the house evokes vernacular tradition, yet its elements are far from representational: the entrance gable is greatly overscaled, as are its two supporting columns. The entry overhang is equally oversized, hovering with no apparent support. The fireplace mass, upon closer inspection, also appears to float above the pitched roof. The south elevation, by contrast, is stripped of all conventional form.

Two-Sided House
双面住宅

上图：从低平宽大的山墙出入口进入室内，首先来到起居室。迎面的是明亮的落地窗，透过窗户可以看到草坪和池塘。可调节的天窗改善了温室的通风。

下图：住宅室内空间的表达是纯粹抽象的，没有什么装饰，惟独有一个线脚，令人想起入口处2英尺(61cm)深的檐口。

右页图：住宅后来扩建了一个车库与一套独立的公寓。

Above: Entering beneath the low, enveloping, superscaled gable, one passes into the living room and confronts a bright, full-height glass wall open to the lawns and pond. An operable skylight provides ventilation for a greenhouse space.

Right: Architectural expression within the house is purely abstract and without ornament, save for one residual molding that recalls the 2-foot (61-cm) deep entrance cornice.

Opposite Page: A later addition to the house comprises a garage with a separate apartment.

Two-Sided House
双面住宅

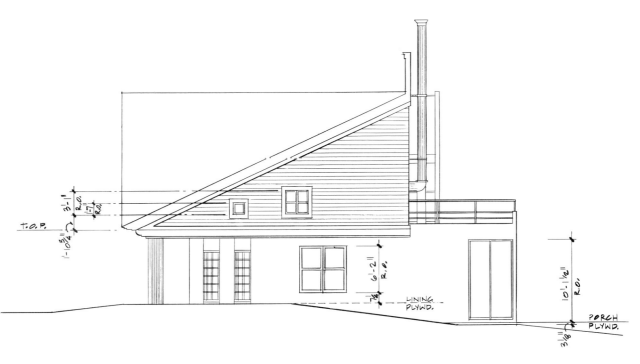

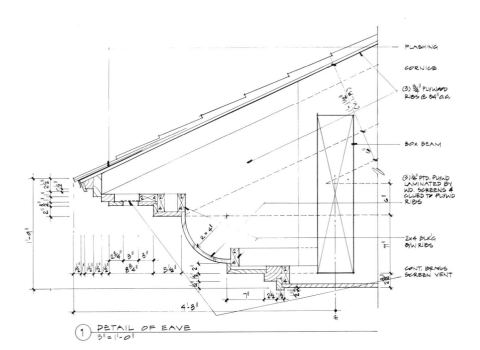

Mies House Pavilions and Second Addition
Weston, Connecticut

密斯住宅的亭和二期扩建
康涅狄格州，韦斯顿

Above: The main elevation of the Mies house was restored but left essentially unaltered by the two new additions.

Opposite Page: The first addition consists of two separate pavilions: one with guest rooms, sauna, and Japanese bath, the other with a small kitchen and party room. The pavilions are linked by a steel screen fabricated by sculptor Richard Heinrich, which also provides a backdrop for the pool and defines the edge of the site.

上图： 密斯住宅的主立面在两个加建的建筑旁边原封未动。

左页图： 第一期加建包括两个亭子：一个亭设置了客人房、桑拿、日本浴室；另一个有公共活动室和厨房。两部分通过钢制的屏风连接在一起，由理查德·亨里奇设计。屏风同时也成为游泳池的背景，界定了地段的边缘。

1955年，密斯·凡·德·罗在其早期的工人住宅基础上，设计了这栋私人住宅。他使用了在建造著名的芝加哥湖滨公寓时剩余的窗墙单元。作为在美国现存的三个密斯住宅中的一个，此住宅在1981年被一名商人买走，他想在扩建了狭小的空间之后用于周末度假。5年后，一位新业主找到建筑师，要求将这栋住宅变成全家永久居住的场所。在第一期工程中，建造了两个亭子，在第二期工程中，将原建筑复原并加以扩建。

给密斯住宅增加一部分的想法极富挑战性。要尊重这个现今已经成为历史的高度现代主义的作品，但不能简单地模仿。因此，出于对密斯和历史的尊重，这个设计应当是语境化的；出于对业主新的要求的满足，在功能上应当比原设计更加完美。所以，在这个设计中，现代主义被当作一种传统，国际风格被看作一种历史模式。

在第一期设计中，原建筑像图标一样原封未动地被搁置在一边，成为整个建筑组合的一部分，另外还有两个分而不离的亭子，这是受到1929年密斯的巴塞罗那亭（Barcelona Pavilion）的影响。在一个亭中，设置了两间客人房、一间桑拿、一间日本浴室；另一个亭中，有一间宽敞的公共活动室和专门为聚会与娱乐而设的厨房。这两个亭通过格子式钢制屏风相联系，同时也界定了室外游泳池的区域。

早期的现代主义者对日本建筑的兴趣也为设计提供了参照。日本风格在一些细部设计上得以体现，如地板升起，室外的墙壁滑动到玻璃匣之中，使得建筑向户外敞开。不过这些日本的暗喻都是字面上的，使用的材料包括高度现代化的玻璃、钢和铝材。以玻璃替代墙体的密斯或许同意将它们全部都摒弃，只不过技术上还不能实现。

Mies House Pavilions and Second Addition
密斯住宅的亭和二期扩建

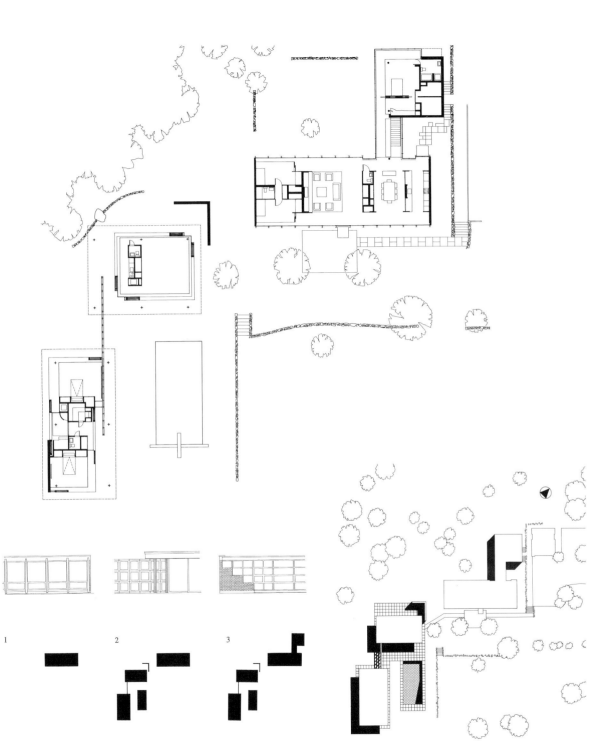

1　2　3

Mies House Pavilions and Second Addition
密斯住宅的亭和二期扩建

上图和左页图：尽管同原建筑分离，但亭的设计体现了密斯的设计理念，并且充分尊重了密斯住宅的整体性。

下图：亭的设计既运用了移动式玻璃也运用了固定玻璃。推拉的玻璃墙可以完全打开；如果需要围合房间时，就在打孔的屏风上镶嵌上玻璃，否则，就保持完全的开敞。

Above and Opposite Page: Although intentionally set apart from the original house, the pavilions produce a composition that both alludes to Miesian theory and respects the integrity of the Mies house itself (right).

Below Left: The pavilions experiment with glass both moving and fixed: sliding glass walls open completely, and the perforated screen is glazed when enclosing a room, and open when out-of-doors.

Mies House Pavilions and Second Addition
密斯住宅的亭和二期扩建

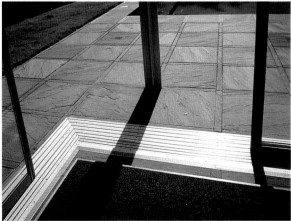

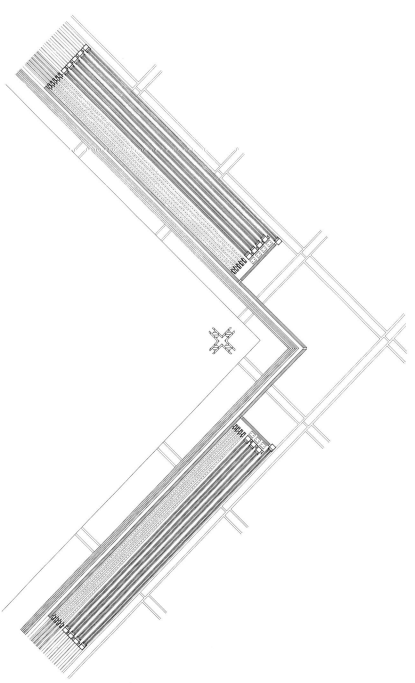

Mies House Pavilions and Second Addition
密斯住宅的亭和二期扩建

左图和左页图：推拉式的玻璃嵌板和隔板储藏在房间对角的玻璃盒中。在空间开敞的亭中，通过下沉天花和抬升地板的方式界定了起居空间。垂直的墙体、水平的屋面在玻璃嵌板上形成朦胧的图案。

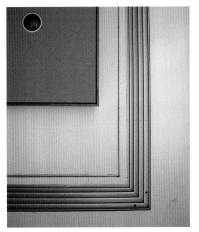

Left and Opposite Page: The sliding glass panels and screens are stored in glass enclosures at opposite corners of the rooms. The dropped ceiling and raised floor demarcate the living spaces within the open pavilions. Reflections of vertical wall and horizontal roof planes on the glass panels create shifting patterns of opaque and transparent images.

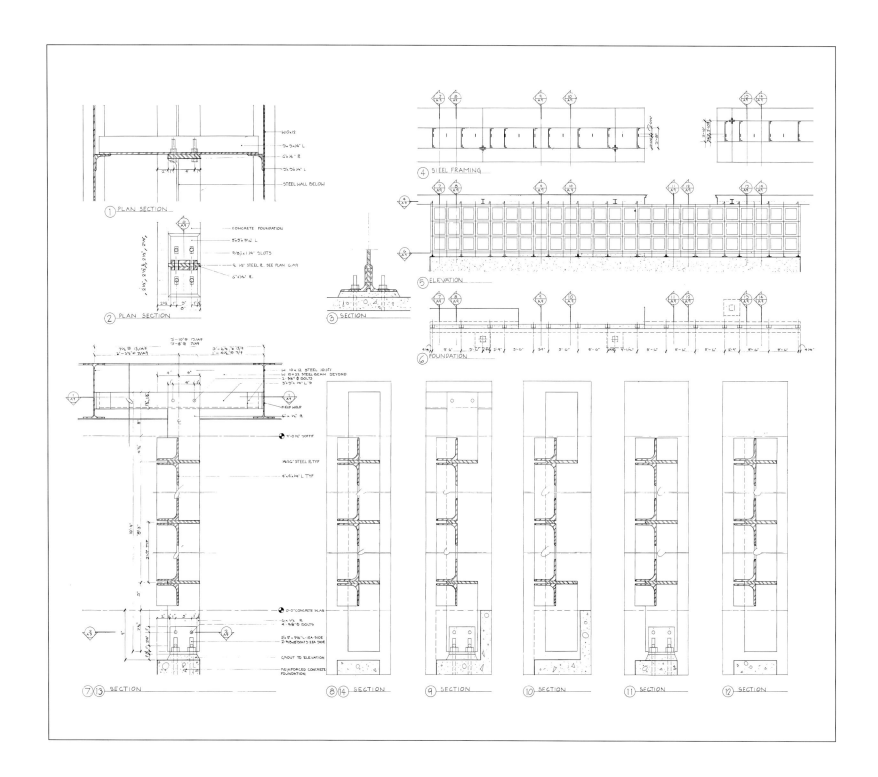

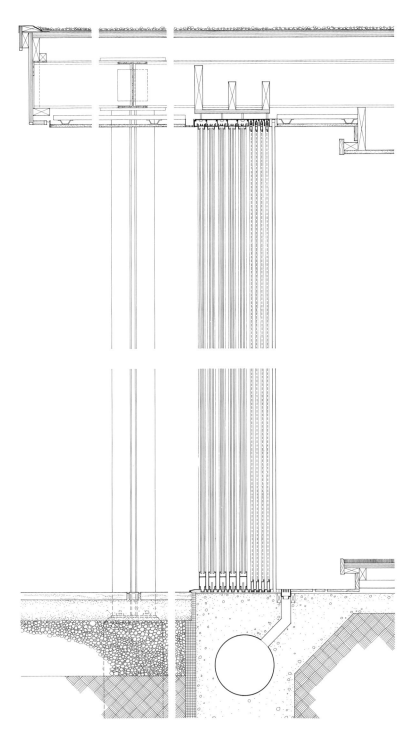

Mies House Pavilions and Second Addition
密斯住宅的亭和二期扩建

Left: *The stone paving patterns, perforated screen wall, and column placement conform to the dimensions of Mies's grid.*

左图：石材铺地、打孔的屏风和柱子的位置都符合密斯式网格的尺度。

Mies House Pavilions and Second Addition
密斯住宅的亭和二期扩建

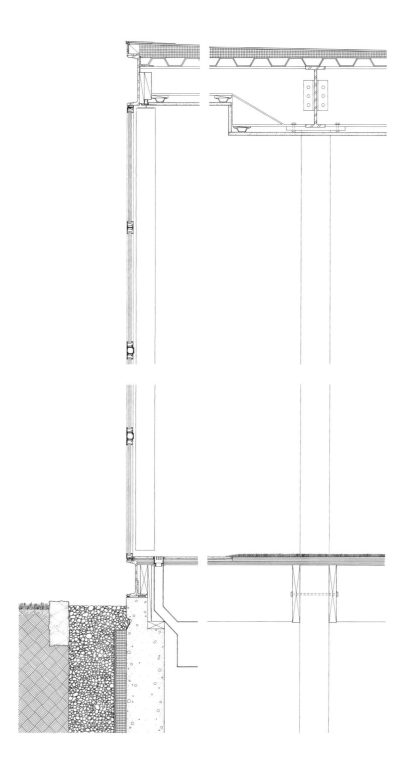

Mies House Pavilions and Second Addition
密斯住宅的亭和二期扩建

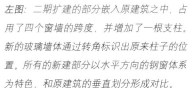

左图：二期扩建的部分嵌入原建筑之中，占用了四个窗墙的跨度，并增加了一根支柱。新的玻璃墙体通过转角标识出原来柱子的位置。所有的新建部分以水平方向的钢窗体系为特色，和原建筑的垂直划分形成对比。

左页图：二期扩建形成入口院落，强调了入口位置。入口占用了一个窗墙开间的跨度，在密斯的原设计中并没有特别的强调。

Left: The second new addition is let into the original structure, replacing four bays of its window wall and inserting one round column for support. The new horizontal glass wall turns the corner and marks the location of the missing Mies columns. All new construction is identified by its horizontal steel window system, which contrasts with the verticality of the original.

Opposite Page: The second addition both creates an entrance courtyard and specifies one of Mies's bays as the entrance to the house, which had not been identifiably marked in the original.

Mies House Pavilions and Second Addition
密斯住宅的亭和二期扩建

48

上图：对起居室的精确复原包括更换腐朽的木制嵌板，铺设原设计的石灰石地面（从未铺过），修理钢制的屋顶系统与窗墙单元。

下图：二期的扩建包括主人卧室，完整的厨房，独立的餐厅，地下娱乐室及其设备。

Above: Accurate restoration of the living room included the replacement of deteriorated wood paneling, installation of the original travertine floor (designed but never laid), and repair of the steel roof system and window walls.

Right: The second addition contains a master bedroom, full kitchen, separate dining room, and basement playroom and utilities.

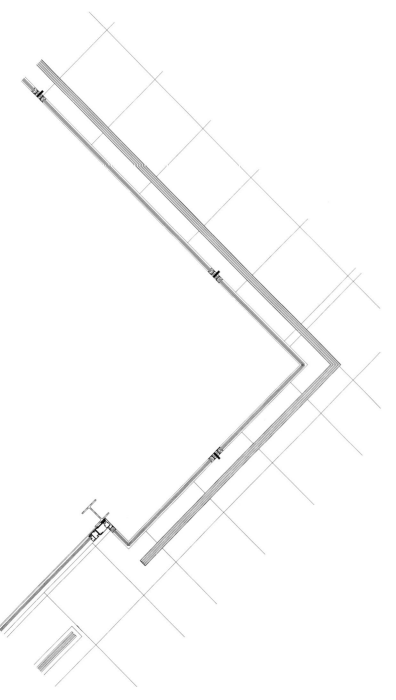

Mies House Pavilions and Second Addition
密斯住宅的亭和二期扩建

上图、中图：在设计中扩展了密斯运用玻璃的手法，将玻璃当作独立的客体对待。图中展示了通往新建卧室和浴室的走廊。

下图：在主人卧室套房的工作区域，一根钢柱支撑着形式自由的枫木写字台。

Above and Center Left: *Extending Mies's use of glass, the design uses glass as an object itself, here shown at corridor to new bedroom and in new bathroom.*

Left: *A steel column (detail) supports a free-form maple desk in the study area of the master bedroom suite.*

Addition to Usonian House

Pleasantville, New York

"美国风"住宅扩建

纽约，普莱森特维尔

1954年，赖特的学徒设计建造了这栋住宅，造价不到10000美元，位于赖特设计的"美国风"居住社区之中。和赖特亲自设计的邻居相比，此住宅不但在设计上有所偏颇，而且由于价廉的建筑材料年久失修，破败不堪。新业主希望能扩大空间，因此要对住宅进行全面的整修。

设计的目标是修正建筑笨拙的一面，使之能够更好的满足更大家庭的需要。设计遇到的挑战是需要改进的比需要保留的多，这样才能使赖特式的要素更加赖特化，使现代风格更加强烈，使建筑与环境的联系更加直接。住宅位于陡峭的林地之上，顶层带有车棚，平面开敞，面向树林景观，这些都十分出色地表达了赖特"美国风"的设计思想。但是建筑的其他部分就不尽人意了，而且底层没有直接对外的出入口。

加建部分是一个筒体，包围了原建筑的下面两层，形成了顶层的底座，使之看起来很像只有一层的建筑，并且以鲜明的赖特手法强调其水平线条。顶层的分隔都去除，代以玻璃，从而提高了形式的纯净性。筒体中有四个卧室和两个浴室，顶部则成为树荫遮蔽下的平台，是非常有用的室外空间，这在以前是没有的。整个住宅有一个楼层用于日常起居和进餐，十分优雅而且实用；卧室都位于下层，儿童室和娱乐室同主卧套间分离。所有的房间都能欣赏树林的美景。

左页图： 从山下看，加建的筒体将原住宅的下面两层围合，形成了顶层的基座。水平方向的暗红色拉毛粉饰围栏、原建筑起居室的屋顶与筒体的白色粉饰形成鲜明对比。

Above: From the street, the soaring roof, carport, and horizontal cypress siding comprise the typical image of Wright's Usonian House. Here it has been preserved, restored, and enhanced with additional glass in the living spaces. From this perspective, the new addition appears only as a platform on which the Wrightian icon sits.

Opposite Page: Seen from the bottom of the hill, the barrel-shaped addition—forming a plinth for the Wrightian top floor—encircles the two lower levels of the original building and contains four new bedrooms. New horizontal railings in dark red stucco, together with the original roof-plane of the living room, contrast dramatically with the white stucco of the new circular form.

上图： 从街道上望去，住宅的屋面如同鸟儿一样展翅翱翔，与车棚、水平的柏木墙板等一起构成了典型的赖特式"美国风"住宅意象。在此，原设计得到了保护、修复，并由于起居空间玻璃的运用，得到了进一步的改善。从这个角度看，加建部分仅仅是一个基座，赖特式的图标位于其上。

52 Addition to Usonian House
"美国风"住宅扩建

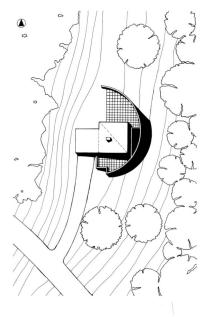
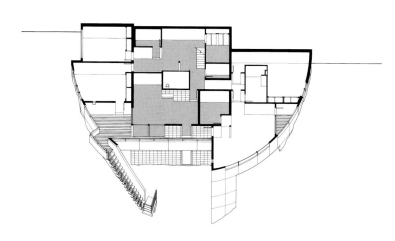

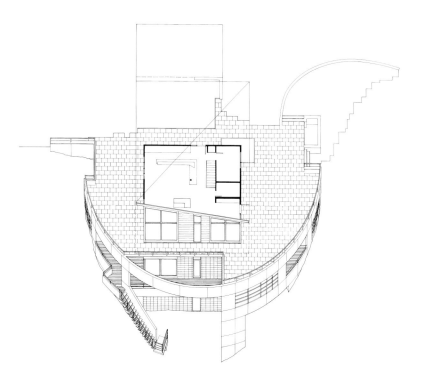

Addition to Usonian House 53
"美国风"住宅扩建

上图: 筒体的屋面既强调了有机的赖特式图标,又为起居和就餐区域提供了宽敞的平台。

左页图: 这个赖特式的设计更适合于缓坡,因此,在这陡峭的地段上显得笨拙。建筑的最大特色在于水平的顶层,但被其他部分的设计破坏了,而且,也没有提供通往树林和山脚下赖特作品的出口。

Above: The roof of the barrel-form both highlights the "organic" Wrightian icon, and provides a generous terrace that surrounds the living and dining area.

Opposite Page: Best suited to gently sloping land, the Wrightian design of the original house seemed awkward on this steep-sided site. The three stories not only compromised the house's best feature—the horizontal upper floor—but there was no access to the wooded landscape or to the common lands of Wright's original site plan, which lie at the bottom of the hill.

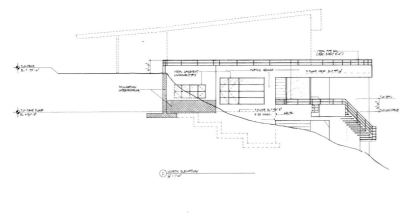
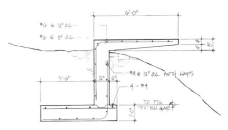

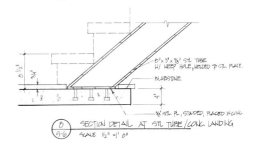
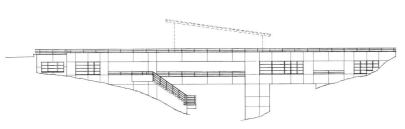

Addition to Usonian House 55

"美国风"住宅扩建

左图： 平缓的楼梯围绕筒体下降到一个混凝土平台，平台与林地有着方便的联系。这个新建的平台和楼梯改善了赖特经常强调的对季节交替的体验。图中，冬雪改变了自然环境和室内的氛围。

Left: A gentle stairway wraps around the barrel, descending to a concrete platform, which permits easy access to the woodlands. The new terrace, deck, and stairway all enhance the experience of seasonal change that Wright so often stressed. Here, winter snows alter the feeling of the natural setting and of the interior of the house as well.

56 Addition to Usonian House
"美国风"住宅扩建

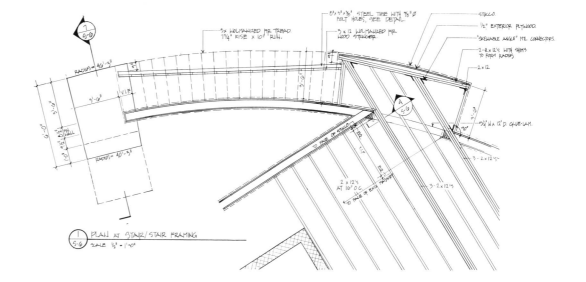

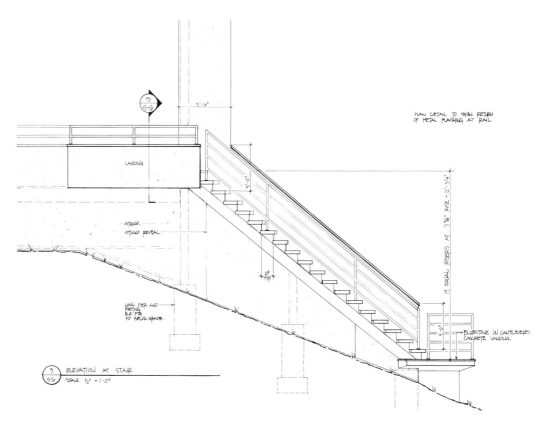

Addition to Usonian House   57
"美国风"住宅扩建

左图：富有雕塑感的镂空设计减小了直径为85英尺（约26m）的筒体的尺度，同时还起到了框景的作用。从外观上看，镂空的部分使原建筑的水平线条时隐时现，而且，筒体本身独具特色，避免了对赖特的直接的参照。

左页图：透过对称的钢窗，从筒体底层的卧室可以欣赏到水平方向的开阔景观。因为地势陡峭的缘故，这些较低的楼层看起来好像悬挂在树林中。

Left: *Sculptural cut-outs reduce the scale of the 85-foot (26-meter) diameter barrel form, and they also frame views of the landscape from inside. From outside, the openings provide glimpses of the horizontality of the original house, while the circular form stands on its own terms, removed from any direct reference to Wright.*

Opposite Page: *Symmetrical steel windows offer horizontal vistas from the bedrooms on the new lower story inside the barrel. Because of the steepness of the site, these lower-level rooms feel as if they are suspended in the midst of the trees.*

58 Addition to Usonian House
"美国风"住宅扩建

右图：室内的改造包括复原木制顶棚和原来颇具雕塑感的混凝土壁炉。长沙发和花盆均是赖特风格的设计。原来不是十分恰当的墨西哥瓷砖被赖特所偏爱的新的纯亚麻地毯取而代之。

右页图：在筒体中挖出一个小阳台，成为主卧独立的户外私密空间。

Right: Interior modifications included restoring the wood ceilings and the original sculptured concrete-block fireplace. A couch and planter were designed in the Wrightian manner, and a new homogeneous linoleum floor—of the sort favored by Wright—replaced inappropriately added Mexican tile.

Opposite Page, Below: A small terrace is cut out of the barrel, giving the master bedroom its own private exterior space.

Addition to Usonian House 59
"美国风"住宅扩建

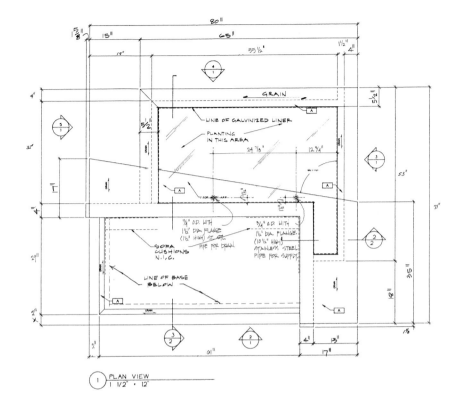
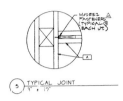
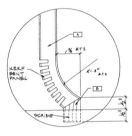
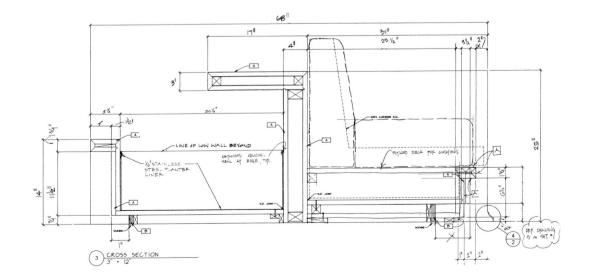

Farmhouse with Lap Pool and Sunken Garden
Worcester, New York

带游泳池和下沉花园的农宅

纽约，沃切斯特

这栋白色外框架的农宅建于18世纪，位于开阔的山地之中，在旁边的小山坡和两座青贮塔的陪伴下，构成了美国乡土建筑的美丽图画。业主要求扩建部分的规模应该是原建筑的两倍，包括一个艺术画廊和一个室内游泳池。

在这样的环境下产生的设计应该是可以多重解读的，不去限制新建筑各种雕塑化形体的可能性，同时应该保持原建筑的主要地位。新的附属建筑以传统的方式布置在老建筑后部，其形状与材料和地段中的谷仓与青贮塔遥相呼应，通过抽象的手法表达了建筑师主张的"语境化"的现代主义。同样的原则在对待现代主义大师密斯和赖特的作品时早已有所发展，运用在这个无名的农宅扩建设计中，通过尊重与对比达到了改善原建筑的目的。

方案包括卧室、主卧套间、画廊和游泳池四个分离的形体，避免了过大的建筑体量。为了在新和旧之间保持适度，15英尺（约4.6m）高、95英尺（约29m）长的游泳池严格限制在一层的高度。因此，游泳池朝向下沉花园开口，保证了一定的私密性，也避免了由于增设围墙或栅栏而对开阔的景观形成的破坏。建筑不仅受到整个环境的影响，其自身也在创造景观，将整栋住宅朝向外部空间扩展。建筑并没有在建筑物的外墙外止步，而是同室外再造过的地坪和绿化组成的表面所创造的空间统一起来。

这样布置既可作为家宅和退休后的休养地，又可作为当地艺术品的陈列中心。这两种功能在空间上象征性地分开，但彼此相隔不远。

Above: The original farm buildings and the shapes of the new structure form a modernist composition that also evokes vernacular rural architecture.

Opposite Page: The sunken garden on the pool level—directly accessible from the gallery and master bedroom—is cut into the flat plane of the river valley site. Because the view is just below ground level, privacy is assured without the need for walls that would obstruct the open landscape.

*上图：*原农宅和新建筑的外形是现代主义风格的组合，但仍然不失乡土情调。

*左页图：*下沉式花园镶嵌在平坦的河谷之中，和游泳池处于同一地平，从画廊和主人卧室可以直接到达。因其低于地表，不需要围墙就能够保证私密性，避免了对大地景观的破坏。

Farmhouse with Lap Pool and Sunken Garden
带游泳池和下沉花园的农宅

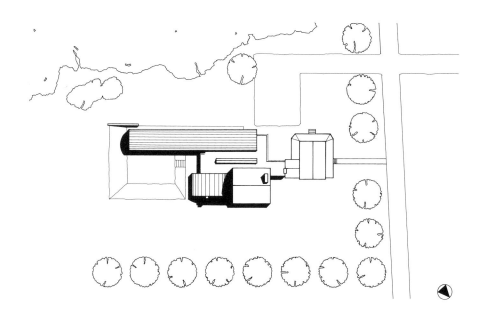

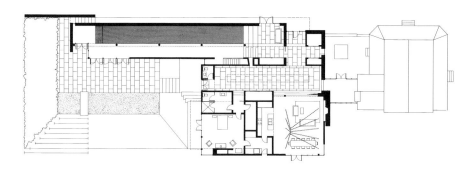

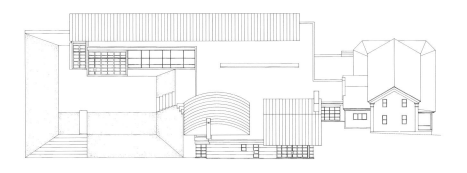

Farmhouse with Lap Pool and Sunken Garden
带游泳池和下沉花园的农宅

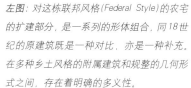

左图：对这栋联邦风格(Federal Style)的农宅的扩建部分，是一系列的形体组合，同18世纪的原建筑既是一种对比，亦是一种补充。在多种乡土风格的附属建筑和规整的几何形式之间，存在着明确的多义性。

左页图：材料的色彩和质感进一步改善了现代风格的形式和形体。游泳池区域使用了多种材料，如杂色的印度石板、褐色拉毛水泥、铝板顶棚、天然石墙以及经过打磨的山毛榉木柱。

Left: The additions to the Federal-style farmhouse are conceived as a composition of shapes that both contrast with and complement the eighteenth-century original. There is conscious ambiguity between the impression of multiple vernacular outbuildings and the reading of formal geometries.

Opposite Page: Color and texture of material enhance modernist shapes and forms. The pool area combines multicolor Indian slate, ochre stucco, aluminum-leaf ceiling, fieldstone wall, and polished natural-beechwood columns.

64   *Farmhouse with Lap Pool and Sunken Garden*
带游泳池和下沉花园的农宅

Right: The lap pool creates a transitional space between the main living areas and the sunken exterior garden. In the summer, the steel-and-glass wall opens to make the pool continuous with the outside. The bar and seating area on the upper level integrates the pool space into the more formal rooms of the house.

右图: 游泳池构成了起居厅和室外下沉花园的过渡空间。玻璃幕墙使室内外交融到一起。在上一层的吧台与休息区将游泳池空间同其他更为正式的房间统一起来。

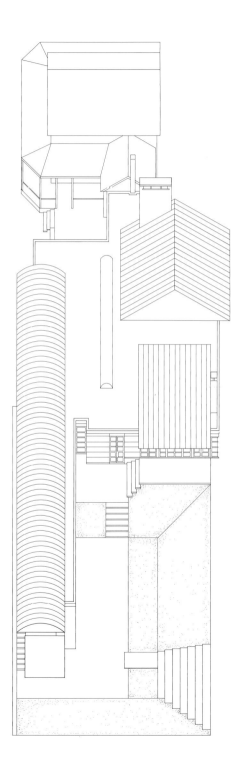

Farmhouse with Lap Pool and Sunken Garden

带游泳池和下沉花园的农宅

Left: The play of light and shadow through the window walls and the patterns created by the different materials enrich the simple shapes, and they enhance the ambiguity between inside and outside.

左图：光线穿过玻璃幕墙形成的变换和不同材料形成的图案，丰富了简洁的形体，进一步模糊了室内外空间的差别。

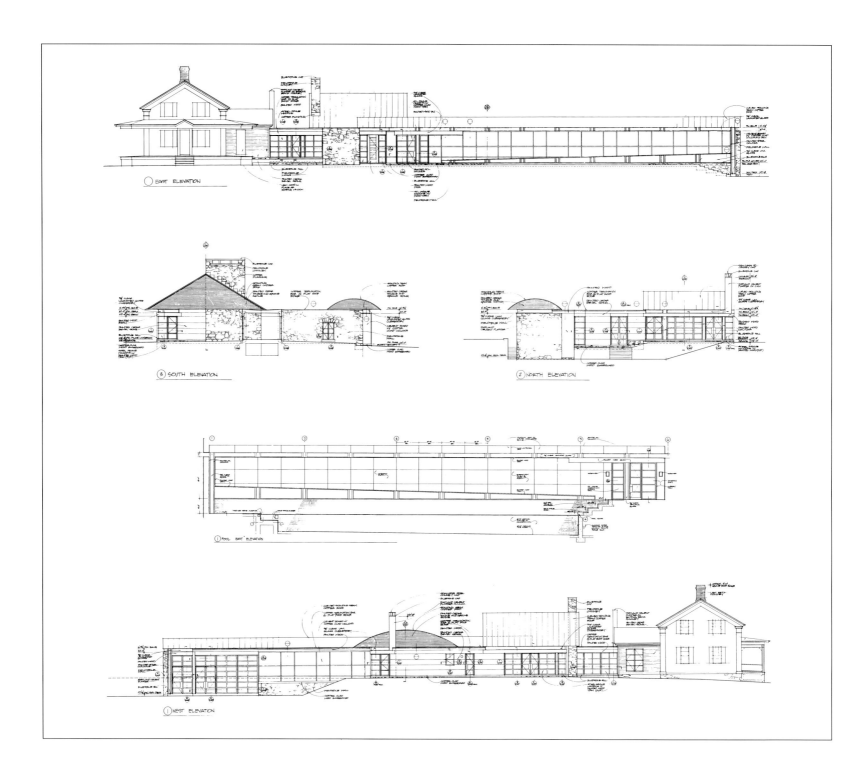

Farmhouse with Lap Pool and Sunken Garden

带游泳池和下沉花园的农宅

右图：支撑屋顶的木制桁架赋予巨大的空间近人的尺度，暗示着这间多用途的房间的区别性。主卧、浴室和画廊室内空间反映出外部的形式，保持了形体的一致性。

右页图：以传统的方式运用不协调的材料，产生了在正式和随意之间的对比：一方面是规整的形体，另一方面是随意的环境。

Right: The structural wooden truss supporting the roof gives human scale to a very large space and implies differentiation within this multiuse room. Reflecting their exterior form, the master bedroom, bathroom, and the hall art gallery retain their formal identities.

Opposite Page: Conventionally inconsistent materials are used to express a contrast between formality and informality: the formal geometry on the one hand and the informal setting on the other.

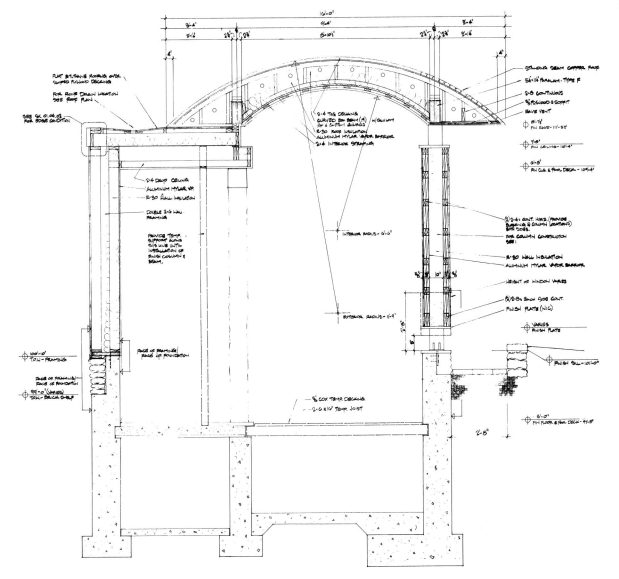

Farmhouse with Lap Pool and Sunken Garden
带游泳池和下沉花园的农宅

Linear House

Millerton, New York

直线型住宅

纽约,米勒顿

Opposite Page and Above: *The roofs of the old farmhouse and the new addition are separated, gable end to gable end, establishing kinship and individual identity at the same time.*

左页图及上图:新旧建筑的屋顶是分离的,山墙相对,既相似又不泯灭个体的存在。

新英格兰的农夫在通往市场与城镇的路的右旁建造自己的住宅。他们的农务活动,如种植蔬菜、养鸡等,都尽可能地与之靠近。因此,一户人家就成为从乡间环境中隔离出的井井有条的小块区域。今天,高速公路不再那么富有吸引力,而令人神往的乡村景观则代表着私密、美景和倘佯其中的惬意。

这栋建于19世纪早期的住宅属于一个有四个孩子的家庭,其扩建方案以20世纪晚期的生态自然观,处理住宅和地段的关系。新建部分80英尺(约24m)长,从路边的原住宅向内延伸,同时在立面上形成通向池塘、瀑布和古老的苹果园的过渡,而这些景观在以前既看不到也无法接近。

这是建筑师"语境化的现代主义"系列中的最新作品,其建筑语境有两个要素:其一是周围景观,其二是普通的乡土风格。建筑以简洁的金属屋面体现了乡村的实用主义。新建的室外走廊利用了地形,从原建筑的二层通向池塘和果园,这在以前曾经是谷仓和干草棚建立方便联系的方式。

新建部分采取了纯粹的直线形式,两侧是悬挂在梁柱结构上的玻璃立面。只有一个房间比较深,其规整的设计可以保证每个房间都能看到瀑布和另一侧的石头公园,同时,交叉式的通风设计节省掉了空调设施。直线型的平面最大程度地保障了私密性,新建的房间并不互相毗连,除了两间卧室、学习室、家庭活动室之外,还有单独的儿童房与客人房。全新的厨房联系着新旧住宅的日常生活功能,整个住宅都是为满足有大量室内外活动的随意的生活模式而设计。

Linear House
直线型住宅

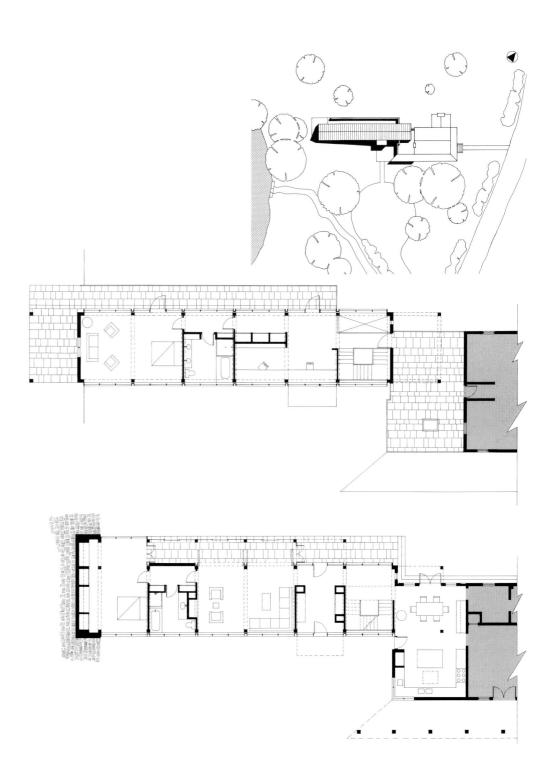

Linear House
直线型住宅

上图： 从田野中望去，新建部分的直线形体和原建筑的特殊形式形成了对比，屋顶采用了一整块长90英尺（约27m）的屋面。

下图： 后立面体现出14英尺（约4m）宽的简洁的平面形式。

Above: *Viewed from across the field, the bold linearity of the addition contrasts with the ad hoc shapes of the original farmhouse. A single 90-foot-long (27-meter-long) roof encloses the entire new program.*

Left: *The rear elevation expresses the simple, 14-foot-wide (4-meter-wide) plan of the addition.*

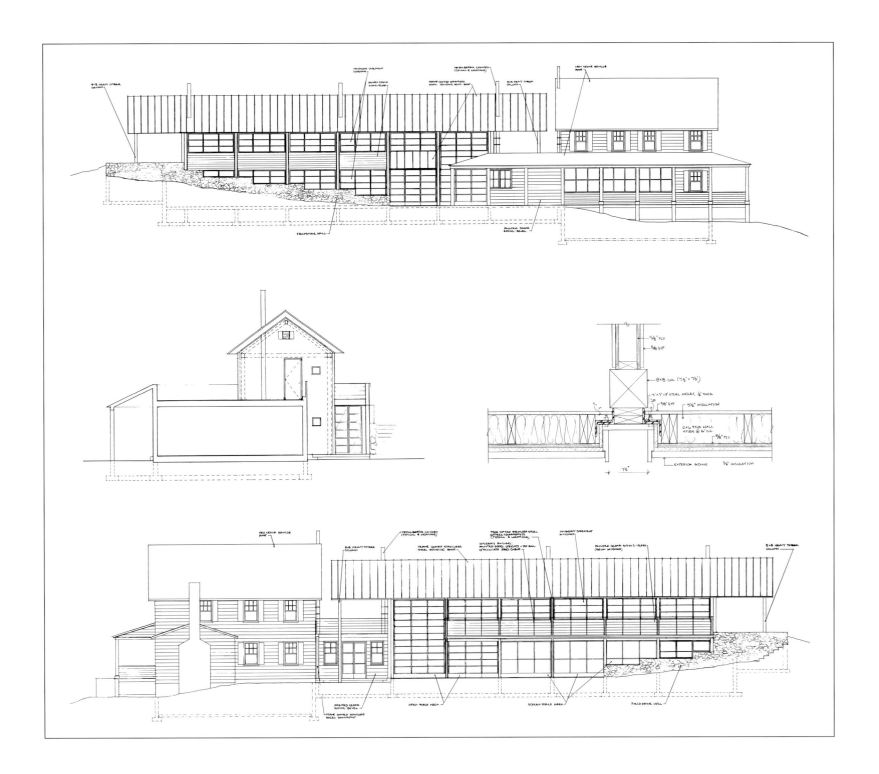

Linear House
直线型住宅

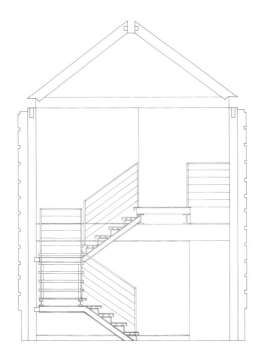

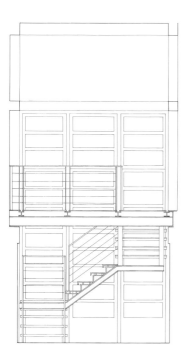

右图：共有8个开间，每个房间占用1到2个。楼梯及其通向原建筑的走道占用了1个开间，在顶层向毗邻的书房开敞，并展示出室内空间的高度。从主卧室中可以毫无遮挡地看到整个建筑的通长及其结构系统的韵律。

Right: Each room occupies the entire width of the wing and either one or two of eight structural bays. The stair with its bridge to the old house uses one bay of both floors, with the top floor open to the adjacent studio space, and reveals the full height of the building. From the master bedroom there is an unobstructed view of the full length of the building, which reveals its repetitive structural system.

Linear House

直线型住宅

Left: *The structural wood frame and its connections are exposed, making reference both to vernacular rural construction and to rational, modernist expression.*

左图：木构架及其连接方式全部暴露在外，乡土风格的构架通过理性的现代主义得以表达。

Bridge House

Olive Bridge, New York

桥式住宅

纽约，奥利夫布里奇

Above: A 90-foot (27-meter) bridge leads from a roof terrace to a high ledge of bluestone, ferns, and typical mountain flora. The house itself is a bridge linking this forest environment to the formal gardens and walkways on the rolling lawns below.

Opposite Page: A formal lawn cut through a wooded landscape terminates at a symmetrical cube. The lighted hallway of the bedroom wing pierces an intersecting vertical form, then disappears obliquely off-axis into the woods.

这栋属于一户城市居民的乡间别墅，重新塑造了随处可见的普通度假住宅的形式和功能。现代主义既没有戴着坡屋顶的面具，也没有披挂上多重解读的手法，它是大胆且直率的，在空间和形体上表达着建筑的现实性。在这个特别的例子中，"桥"是地段和设计在概念和物质空间上的需要：这就是位于历史的形式、风格和材料之间的"桥"，位于地形、使用和几代人之间的"桥"。

住宅由三个明确的形体构成，每个都有自己的风格和独特的材料。主立面是巨大的三层筒体，表面是如同石板一样的薄薄的混凝土板，使人想起早期美国建筑所拘泥的帕拉蒂奥式风格。两侧有檐板的陡峭的坡屋顶与筒体斜交，呼应了当地农舍的形式。一座长长的、狭窄且被架高的侧翼穿过其中，外挂金属波纹板，这是20世纪低造价乡间住宅的典型方式。整栋住宅令人想起过去三个世纪的美国乡村建筑。

在筒体的屋顶平台之上，有一座长长的桥通向青石山崖，将规整的草坪、下方的玫瑰花园和岩石、瀑布以及旁边的树林联系起来。室内的中心是筒体宽敞的起居室中的公共区域，其他次要一些的空间是位于坡屋顶下面的图书、台球室和学习室，五个卧室位于直线型的侧翼之中，以保证私密。每间卧室都面向走廊，有俯视树林的窗户，其间插入的浴室起到了较好的隔声效果。几代人都住在这栋住宅中，因此设计为不同的年龄和家庭提供了不同的空间，总共可以容纳24个人。

室内和室外空间都充满了奇思妙想，因此空间的感觉多种多样且富有趣味，这是在树林中居住和娱乐的难得去处。

上图：90英尺（约27m）长的桥从屋顶平台通向高高的青石岩架、蕨类和其他典型的山地植物群落。住宅本身就是一个桥梁，联系树林环境同规整的花园以及起伏的草坪之下的步道。

左页图：规整的草坪嵌于树林景观之中，一直延伸到对称的筒体之下。卧室侧翼明亮的走廊穿过相交的垂直形体，然后偏离轴线，消失在树林中。

Bridge House
桥式住宅

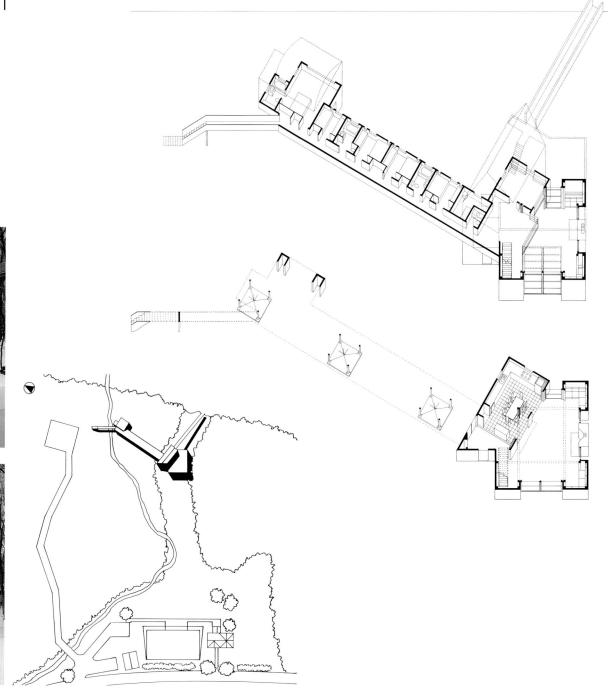

Bridge House
桥式住宅

左图： 架高的卧室侧翼浮于树林景观之中。高高的桥使屋顶和山顶建立了直接的联系，夏天有山间小径，冬天可以沿滑雪道方便地上下。

左页图： 三个相交的形体表达了这样一个构想：混凝土板覆面的筒体住宅成为戏剧性的公共起居和用餐空间；85英尺(约26m)长的金属墙面的侧翼包含着卧室；总共三层的坡屋顶空间容纳了图书室、台球室、学习室、厨房和入口门厅。

Above: The elevated bedroom wing is raised above the wooded landscape. The high bridge gives direct access from the roof to the ridge with its summer hiking paths and winter cross-country ski trails.

Opposite Page: The three intersecting forms express the program: the concrete-board-clad cube houses the dramatic public living and dining spaces; the 85-foot (26-meter) long metal-faced wing contains the bedrooms; and the three-story pitched-roof encloses the library, billiard room, study, kitchen, and entry hall.

Bridge House
桥式住宅

Bridge House

桥式住宅

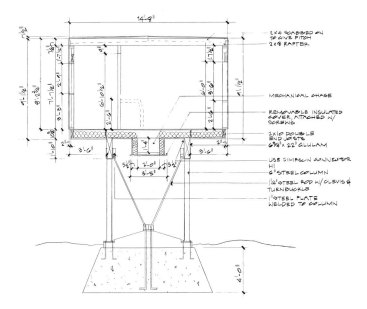

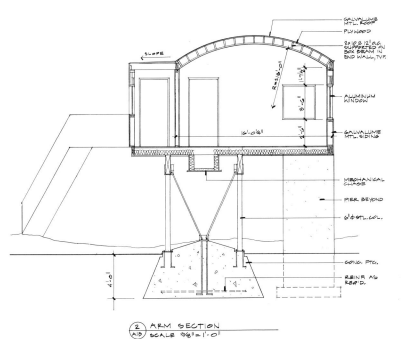

Opposite Page and Left: *The single-story bedroom wing was designed to make the least impact on the natural landscape while providing the greatest privacy for guests. Insulated from one another by bathrooms and closets, the bedrooms are cozy, private retreats removed from the social action in the main parts of the house.*

左页图及左图：只有一层的卧室侧翼在设计上尽量减小了对自然环境的影响，同时也为客人提供了最大限度的私密。卧室由浴室和卫生间相互隔开，尤其显得温暖、舒适和安静，隔绝了住宅主体中公共活动的影响。

Bridge House
桥式住宅

Right: The seven-foot high fireplace is framed within an 18-foot-high (5-meter-high) window wall. Windows operate as oversized, double-hung units, with lower sections counter-weighted by stacks of bricks. With windows fully raised, the house becomes a pavilion, entirely open to the outdoors.

右图: 7英尺(约2.1m)高的壁炉嵌入18英尺(约5m)高的落地窗之中。窗户是超大尺度的双扇吊窗,下面的部分由砖垛平衡重量。当窗户全部打开时,房间就像帐篷一样,向室外完全地开敞。

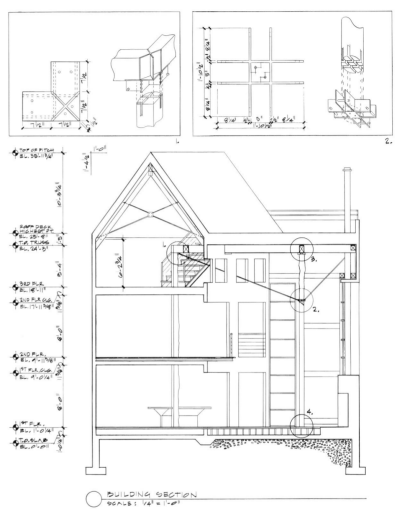

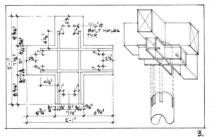

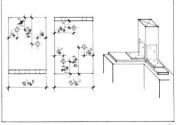

Bridge House
桥式住宅

Left: *A series of inverted, king post trusses spans the large living room and supports the bluestone-paved roof terrace above. Steel stairs and balconies lead to the library, billiard room, and bedroom wing on the second floor, to the study on the third floor, and up to the roof terrace and bridge on the fourth level.*

左图：一系列倒置的中柱式桁架跨过巨大的起居空间，支撑着上面青石铺成的屋顶平台。钢制楼梯和阳台通向图书室、台球室和位于二层的卧室侧翼以及三层的学习室、屋顶平台和四层的桥。

Bridge House
桥式住宅

右图: 大型的带形窗贯通整个走廊,为每间卧室提供了交叉式的通风以及树林开敞的景观。客人卧室总共有一个大间、四个小间,每个房间内都有写字台、工作空间、书架、抽屉、卫生间、行李贮藏间和浴室。内置式双层床能从墙面上放下并展开,提供更多的床位。

右页图: 在大多数房间中均能看到桥,强调了住宅和环境的整体感。

Right: A large strip window runs the full length of the bedroom corridor, providing cross-ventilation for each room and opening the hallway to views of the woods. One large and four small guest bedrooms each have a desk and work space, bookshelves, drawers, a closet, storage space for luggage, and a bathroom. Built-in double-decker bunks fold down from the wall for extra sleeping accommodations.

Opposite Page: The bridge can be seen from most rooms, emphasizing the integration of house and landscape.

Bridge House
桥式住宅

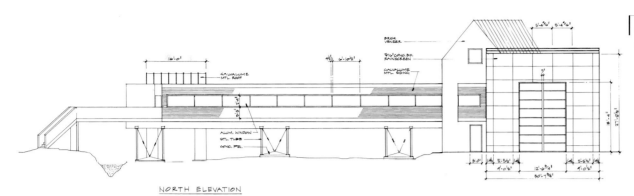
NORTH ELEVATION

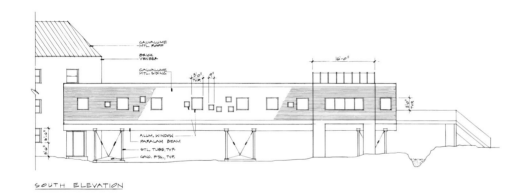
SOUTH ELEVATION

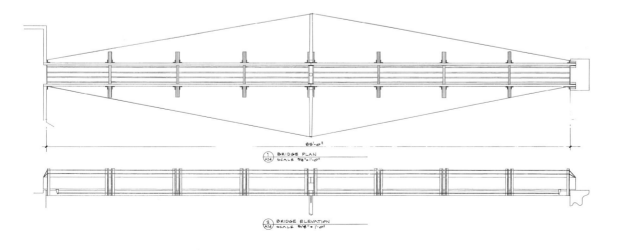
BRIDGE PLAN
BRIDGE ELEVATION

Lake House with Court

Highland Park, Illinois

有内院的湖畔住宅

伊利诺伊州,海兰帕克

Above: The circular-entry courtyard is faced by the segmented steel windows of the family room. The deep zinc roof overhang provides shelter for the exterior walkway and interior shade from the summer sun. The roof is also an element in the landscape design, focusing sheets of rain, hanging icicles, and piles of snow along the curve of its edge.

Opposite Page: From the residential street, what is in fact a very large house appears in an intentionally low and modest outline. (Photos show house under construction)

上图: 圆形内院的立面由家庭起居室中分隔的钢窗组成,锌板屋面的深深挑檐形成了室外走廊,也为室内提供了夏日遮荫。屋面本身也是景观设计的重要元素,比如,下雨时形成的水帘,冬日沿屋面曲线形成的积雪与冰挂。

左页图: 从居住区道路上看去,本来十分巨大的建筑,经过设计只展露出低矮、谦逊的外轮廓。

这栋住宅坐落在鸟瞰密歇根湖的山崖之上,应和着大湖的美丽和惊涛骇浪,创造了两个截然不同的世界:一个面对着日出、湖水和变换的景观,一个面对内向式的院落享受着阳光的温暖,同经常恶劣的湖畔气候形成对比。

建筑强烈的形体也表达出两个世界:曲线式的内院立面从面向大湖的三层的玻璃长方体上撕扯开来。重要的房间,比如起居室、餐厅、家庭室、主卧室、办公室和门厅既朝向内院也朝向大湖。三英尺厚的砖墙从住宅中弯曲而过,是一种人工与自然景色相分离的现代主义暗喻。

住宅抛弃了传统郊区住宅的模式,那种模式将前院、前立面和车库都面向街道。在此方案中,体量巨大的建筑(12500平方英尺,约1125m²)只有很少一部分暴露在社区中。圆形的内院以一堵墙和街道隔开,私密性的室外景观替代了通常开敞和无用的前院。汽车从弯曲的砖墙一侧可以直接开进车库而不被发现。

住宅被设计成富有层次的序列,形成了一系列的景观,季节与气候的变化如同日常生活的行为一样成为建筑体验的一部分。平面设计考虑了大型的娱乐和家庭聚会,同时也可以成为一对夫妇的舒适寓所。

94 *Lake House with Court*
有内院的湖畔住宅

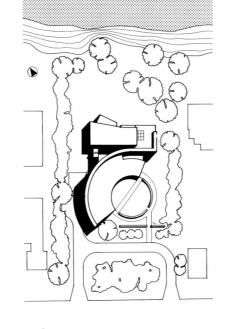

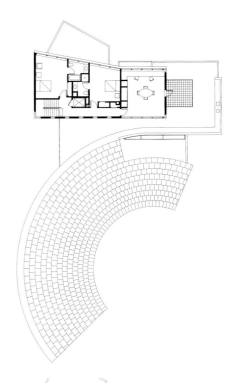

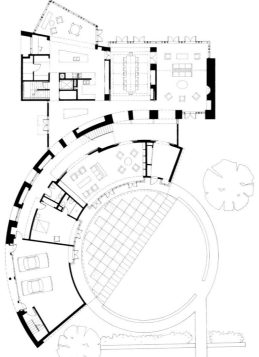

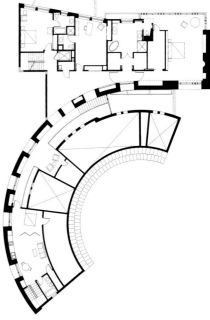

Lake House with Court

有内院的湖畔住宅

左图: 透过双层玻璃窗,即使在内院一侧也能看到密歇根湖的美景,那么可以想像从室内欣赏到景色是怎样的富有戏剧性。

左页图: 早期设计的模型展示了圆和方的形体组合关系。

Left: A two-story glass opening offers a view of Lake Michigan even from the courtyard side, anticipating the dramatic vistas that open up once inside the house.

Opposite Page: This early designed model shows the relationship between the circular and orthogonal composition of forms.

Lake House with Court
有内院的湖畔住宅

Right: The compound curve of the roof is clad with thirteen rows of specially fabricated zinc shingles. The circular courtyard is crossed by a walkway that separates an abstract checkerboard garden of eroded stone and moss from a grass semicircle beyond.

右图：屋顶的复合曲线上覆盖着13排专门生产的锌板。圆形的内院被一条步道穿过，一侧是由已经侵蚀的石材和苔藓形成的花园，其铺地为抽象的方格图案。另一侧是半圆形的草坪。

Lake House with Court
有内院的湖畔住宅

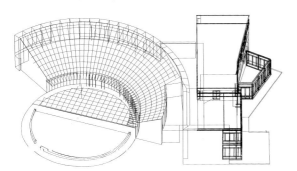

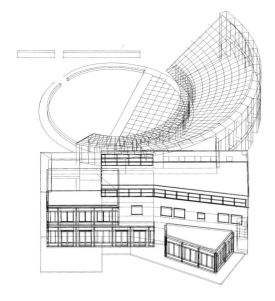

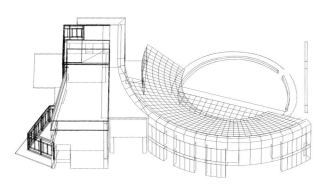

Left: A three-foot-thick brick wall runs through the house, separating the orthogonal from the circular form and the rooms overlooking the natural setting of the lake from those facing the formal courtyard. As the two shapes split apart, they create an entrance on one side and light for the interior hallway on the other.

左图：3英尺(约1m)厚的砖墙穿过住宅，将长方形和圆形的部分分开，鸟瞰大湖的房间与朝向院落的房间亦然。两个分离的部分在一侧形成入口，另一侧为室内提供采光。

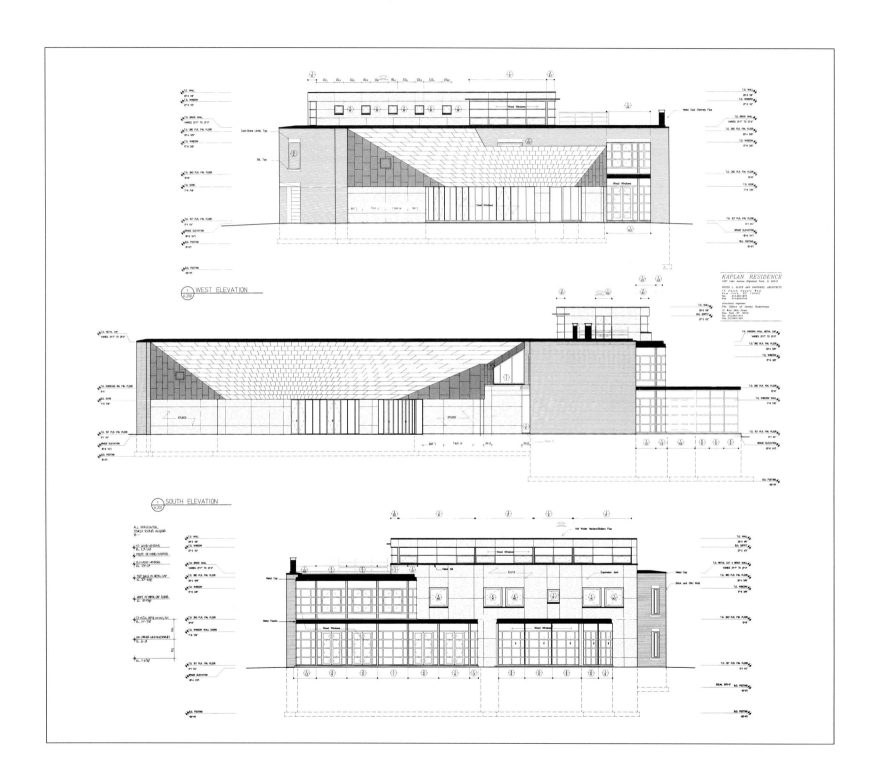

Lake House with Court

有内院的湖畔住宅

Left: The rear facade incorporates a variety of window types that capture the changing scenery of the lake: a two-story glass curtain wall, linear and square-punched windows, and a projecting glass-enclosed room whose window walls drop below floor level to create an open summer porch.

左图：后立面组合了各种不同的窗户类型，从而获得湖面不断变化的景色，二层高的玻璃幕墙，线性和四方形镶嵌的窗户，以及窗户墙下垂到一层的突出的玻璃环绕的房间从而形成了一个夏季走廊。

Urban House

Brooklyn, New York

城市住宅

纽约，布鲁克林

在建成的城市居住区中，设计建造一栋大型住宅，既要体现独特的现代风格，又不能破坏居住区的肌理。而且，对于转角处的地段，要特别注意建筑面向街道的两个立面。

业主是来自欧洲的大家庭，喜欢居住在城市风格的密集型社区之中，在住宅里需要安排11间卧室和12间浴室。方案含蓄地表达了欧式风格，尤其是强调简洁和形体力度的早期维也纳现代主义。

整个建筑为三层半高的立方体，但分割成三个小体量的组合，每一部分的房间都较小，以此避免过于雄伟的立面造成和周围环境的不协调。很多位于街道拐角的住宅均有立面和侧立面之分，而这座建筑不同，转角部分是连续且具雕塑感的形体，在两侧的立面上以等同的方式结束。第三层在转角处内凹形成平台，由花架和屋顶花园围合，沿街的美国梧桐在其上形成了天然的帐篷。

房间围绕着中央楼梯和大型天窗组织。因此，当遮挡沿街窗户以保证私密性时，可以从建筑中部采光。空间的公私分离对业主很重要，所以在门厅里设置了两层高的木框玻璃屏风，以遮挡主要的楼梯。简洁的现代式细部采用了丰富的材料，包括红木、彩色灰泥和石材。三层是钢架和玻璃围合而成的阳光室，朝向室外平台开口。室外平台用于种植花木、娱乐和儿童玩耍，这样的设计在这个城市中颇具匠心。

Opposite Page and Above: *The cubic mass fills the urban site, but is broken into three parts, both to "turn the corner" and to reduce the scale of a structure much larger than its neighbors. The corner, which faces south, is further articulated by an exterior terrace carved out of the third floor, with a deep trellis at roof-height that implies the completion of the cube. The hundred-year-old trees along the street are incorporated into the design to provide an Arcadian bower shading the elevated private terrace.*

左页图及上图：正方体占满了整个地段，但分裂成三个部分，以强调"转角"和减小体量。朝南的转角通过三层的平台得以进一步强调，上面和屋面等高的花架暗示了正方体形的完整。沿街的百年老树也融入设计之中，高高的私密的平台位于天然的遮阳伞之下。

Urban House
城市住宅

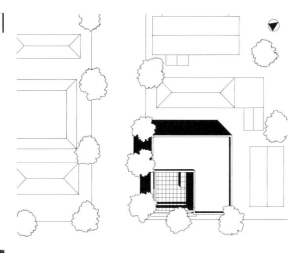

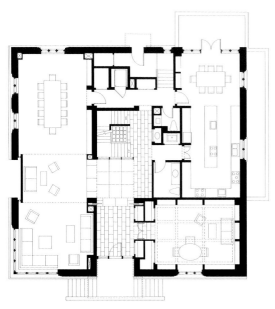
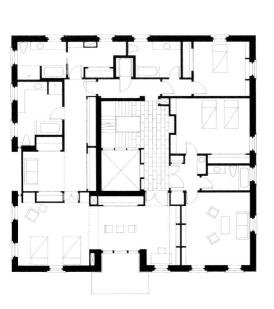

Urban House

城市住宅

上图：模型剖面展示了有天窗的入口大厅和玻璃围合的楼梯，它们是室内空间的组织核心，也为60英尺×55英尺(约18m×17m)的平面区域提供了采光。

左下图和左页图：侧面稍向外倾斜的石材基座将主要的起居层抬起，避免了大街上的嘈杂、活动和纷乱的干扰。建筑的三个部分由后退的玻璃幕墙和打磨的不锈钢板分开。宽大的窗户简化了立面。

Above: The model cutaway shows the skylit entrance hall and glass-enclosed staircase, which establish the internal organization of the house and provide natural light to the interior of the 60 foot x 55 foot (18 meter x 17 meter) floor plan.

Below Left and Opposite Page: A slightly canted stone base elevates the main living floor above the noise, activity, and disturbances of the street. Large windows simplify the facades of the three parts, which are separated by a recessed curtain-wall of glass and sandblasted, stainless-steel panels.

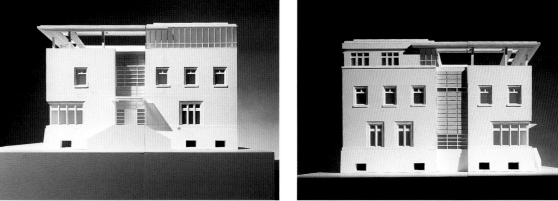

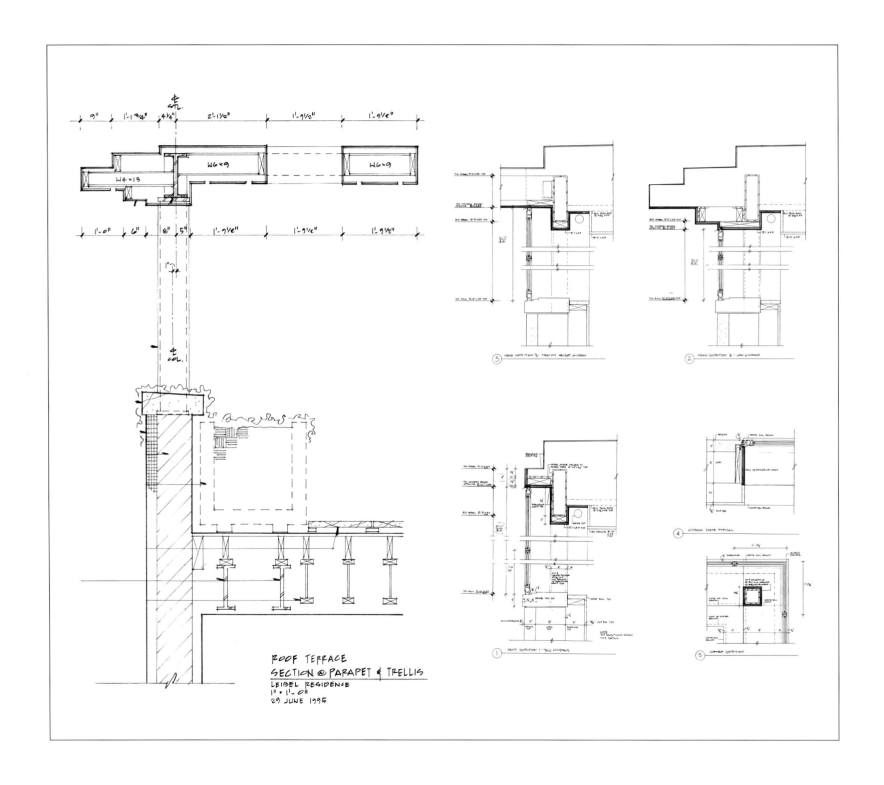

ROOF TERRACE
SECTION @ PARAPET & TRELLIS
LEIBEL RESIDENCE
1" = 1'-0"
29 JUNE 1995

Urban House 105
城市住宅

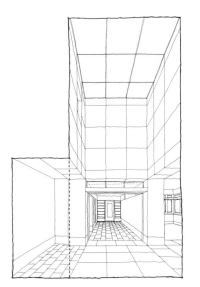
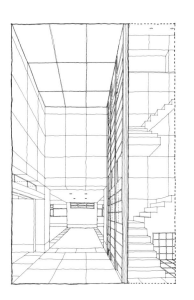
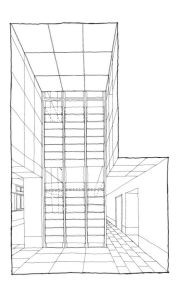

Selected Buildings and Projects
精选建筑与项目简介

"美国风"住宅扩建
纽约，普莱森特维尔

项目助手：温迪·波兹
地段面积：1.25英亩(约0.5ha)
建筑面积：3500平方英尺(约315m²)
设计日期：1992年
竣工日期：1994年

桥式住宅
纽约，奥利夫布里奇

项目助手：托马斯·格卢克
地段面积：18英亩(约7.3ha)
建筑面积：5800平方英尺(约522m²)
设计日期：1992年
竣工日期：1996年

三个山墙住宅
康涅狄格州，莱克维尔

项目助手：迈克尔·马丁
地段面积：2英亩(约0.8ha)
建筑面积：4800平方英尺(约432m²)
设计日期：1985年
竣工日期：1989年

双面住宅
马萨诸塞州，林肯

地段面积：5.4英亩(约2.2ha)
建筑面积：6500平方英尺(约585m²)
设计日期：1980年
竣工日期：1982年

带游泳池和下沉花园的农宅
纽约，沃切斯特

项目助手：弗里茨·里德、基姆·沃克
地段面积：500英亩(约200ha)
建筑面积：5000平方英尺(约450m²)
设计日期：1992年
竣工日期：1995年

有内院的湖畔住宅
伊利诺伊州，海兰帕克

项目助手：温迪·波兹、舒基·狄克逊、弗雷德·沃尔夫
地段面积：1.3英亩(约0.5ha)
建筑面积：12800平方英尺(约1152m²)
设计日期：1993年
竣工日期：在建

马诺尔音乐庄园
纽约，马马罗内克

项目助手：肯特·拉森(合伙人)、卡里·K·戴维斯、马克·海杜克 凯尔文·奥诺、小路光田
地段面积：4英亩(约1.6ha)
建筑面积：11800平方英尺(约1062m²)
设计日期：1985年
竣工日期：1989年

密斯住宅的亭和二期扩建
康涅狄格州，韦斯顿

亭
项目助手：肯特·拉森(合伙人)、路易斯·特平
地段面积：5.5英亩(约2.2ha)
建筑面积：2050平方英尺(约184.5m²)
设计日期：1981年
竣工日期：1986年
二期扩建
项目助手：肯特·拉森(合伙人)、卡里·戴维斯
建筑面积：1500平方英尺(约139m²)
设计日期：1985年
竣工日期：1989年

直线型住宅
纽约，米勒顿

项目助手：舒基·狄克逊
地段面积：50英亩(约20ha)
建筑面积：2500平方英尺(约225m²)
设计日期：1993年
竣工日期：1996年

城市住宅
纽约，布鲁克林

项目助手：舒基·狄克逊、克雷格·格雷伯
地段面积：5853平方英尺(约527m²)
建筑面积：12000平方英尺(约1080m²)
设计日期：1993年
竣工日期：在建

本书插图由埃利卡·马尼格尔德绘制

Firm Profile
公司概况

彼得·L·格卢克获得耶鲁大学建筑学学士学位,并于1965年取得该校艺术与建筑学院的建筑学硕士学位,曾经在纽约、纽方德兰德等地设计了一系列的住宅,随后去东京为一家一流的日本建设集团设计大型项目。这段经历深化了他对日本传统美学的体验,日本将施工与设计整合起来的高效率的现代方法也对他产生了影响,这些都体现在格卢克后来的作品中。

彼得·L·格卢克合伙人事务所从1972年起在纽约开始了自己的业务,以其设计的完整性和对建筑形体与环境关系的敏感性逐渐树立了声望。公司在全美都有作品,包括旅馆、学校、大学建筑、教堂、住宅、室内和历史建筑的复原等。

格卢克的信条是建筑师必须对从概念到施工的建筑过程负责,这使他能承担设计各个方面的疏漏。在这种信仰的驱使下,他建立了建筑施工服务有限公司(ARCS),拥有从设计到施工管理的全套系统,给客户提供严谨的设计,优质的施工以及在越发困难的建设环境中的费用管理。

格卢克的获奖作品展已经在美国和日本举办过,在全世界都有一定的知名度。他还在哥伦比亚和耶鲁大学的建筑学院任教。此外,他也参与博物馆展览,比如1976年在现代艺术博物馆的关于日本城市化的展览——可感知的城市,1996年米兰三年展中的全球化和地方主义等。

Photographic Credits
照片提供者

Henry Bowles

Mies House Pavilions and Second Addition; p. 40 (lower right); p. 41 (top)

Carla Breeze

Farmhouse with Lap Pool and Sunken Garden; p. 69

Jeff Goldberg, ESTO Photo

Manor House with Music; p. 28 (top and left)

Peter Gluck

Urban House (in-progress photos)

David Macleod Joseph

Urban House (model shots)

Norman McGrath

Three-Gable House
Manor House with Music
Two-Sided House
Mies House Pavilions and Second Addition; pp. 7, 36, 38, 39, 40, 42, 43, 45
Addition to Ursonian House

Barry Rustin

Lake House with Court

Paul Warchol

Mies House Pavilions and Second Addition; p. 37, 46–49
Farmhouse with Lap Pool and Sunken Garden
Linear House
Bridge House